discovermore
Earth Science

Rivers, Lakes, and Oceans

Theia Lake

Britannica Educational Publishing
IN ASSOCIATION WITH
THE ROSEN PUBLISHING GROUP

Published in 2026 by Britannica Educational Publishing (a trademark of Encyclopædia Britannica, Inc.) in association with The Rosen Publishing Group, Inc.
2544 Clinton Street, Buffalo, NY 14224

Distributed exclusively by Rosen Publishing.
To see additional Britannica Educational Publishing titles, go to rosenpublishing.com.

Editor: Danielle Haynes
Book Design: Michael Flynn

Photo Credits: Cover Iakov Kalinin/Shutterstock.com; (series background) Dai Yim/Shutterstock.com; p. 4 https://commons.wikimedia.org/wiki/File:The_Earth_seen_from_Apollo_17.jpg; p. 5 yyuuuyu/Shutterstock.com; p. 7 (top) Emre Akkoyun/Shutterstock.com; p. 7 (bottom) EWY Media/Shutterstock.com; p. 8 ChiccoDodiFC/Shutterstock.com; p. 9 romeovip_md/Shutterstock.com; p. 10 Curioso.Photography/Shutterstock.com; p. 11 Lukassek/Shutterstock.com; p. 13 (top) treiff/Shutterstock.com; p. 13 (bottom) PhotosByMike/Shutterstock.com; p. 14 Kichigin/Shutterstock.com; p. 15 Eric Broder Van Dyke/Shutterstock.com; p. 17 Sergii Figurnyi/Shutterstock.com; p. 18 Designua/Shutterstock.com; p. 19 FloridaStock/Shutterstock.com; p. 20 Tomas Kulaja/Shutterstock.com; p. 21 lavizzara/Shutterstock.com; p. 22 Siberian Art/Shutterstock.com; p. 23 Multishooter/Shutterstock.com; p. 24 Damsea/Shutterstock.com; p. 25 LukaSkywalker/Shutterstock.com; p. 26 Vastram/Shutterstock.com; p. 27 Rich Carey/Shutterstock.com; p. 28 Michael Lingberg/Shutterstock.com; p. 29 SantiPhotoSS/Shutterstock.com.

Cataloging-in-Publication Data

Names: Lake, Theia.
Title: Rivers, lakes, and oceans / Theia Lake.
Description: New York : Britannica Educational Publishing, in association with Rosen Educational Services, 2026. | Series: Discover more: earth science | Includes glossary and index.
Identifiers: ISBN 9781641904254 (library bound) | ISBN 9781641904247 (pbk) | ISBN 9781641904261 (ebook)
Subjects: LCSH: Oceanography--Juvenile literature. | Hydrology--Juvenile literature.
Classification: LCC GC21.5 L35 2026 | DDC 551.48--dc23

Manufactured in the United States of America

CPSIA Compliance Information: Batch #CSBRIT26. For further information contact Rosen Publishing at 1-800-237-9932.

Contents

Blue Marble

Many people like to call Earth the Blue Planet. Others, the Blue Marble. From space, it looks mostly blue with brown patches. That is because almost three-quarters of Earth's surface is covered in water. Less than 1 percent of this water is fresh water that we can drink and use. The rest is mostly salt water in the oceans, and some is frozen at the North and South Poles.

This is one of the most famous photographs of Earth, called *The Blue Marble*. Astronauts from the Apollo 17 trip to the moon took the photo in 1972.

Of all the water on Earth, about 97 percent is in the oceans.

We cannot drink seawater because it has so much salt in it. Instead of quenching our thirst, seawater would dry us out. The body tries to get rid of salt by making us produce more urine. If the body gets rid of too much water through urine without being replaced, it becomes dehydrated.

Consider This

While you read this book, think about how rivers, lakes, and oceans are connected.

Running Rivers

Oceans may be the largest bodies of water on Earth, but rivers are just as important. These are large areas of fresh water that move from one place to another. Rivers flow toward oceans, lakes, or other rivers. Some are full of water all year round, but rivers in dry countries may flow only at certain times of year. Some rivers are only a few miles long. Others, such as the Nile in Africa and the Amazon in South America, are so long they flow through many countries.

Rivers are very important to people. In the earliest times, people made their homes alongside rivers so they had fresh water to drink, to wash in, and to cook with. They also caught fish to eat from the river. Rivers are also important for traveling to different places and trading goods.

The Nile River in Egypt is the longest river in the world at 4,132 miles (6,650 km) in length. It's so large it can be seen easily from space.

compare**and**contrast

How do the rivers pictured in this book look alike? How do they look different?

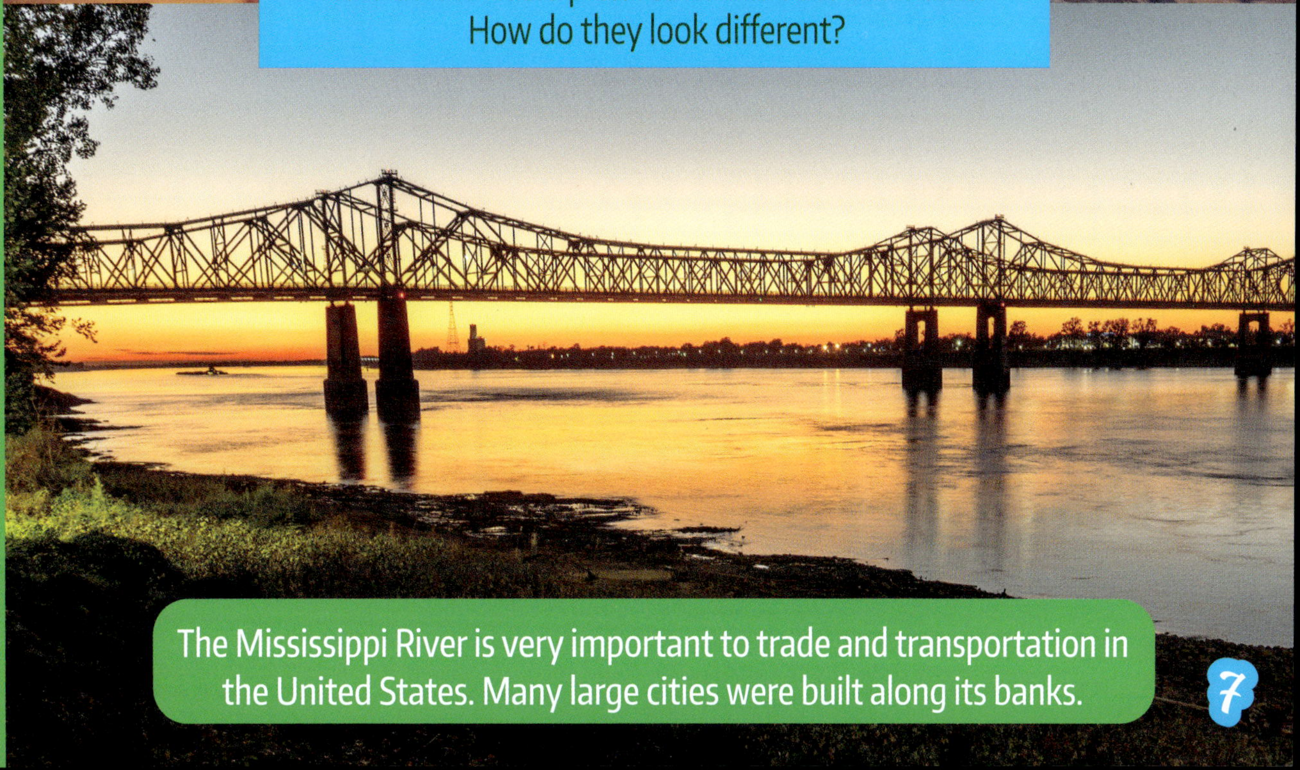

The Mississippi River is very important to trade and transportation in the United States. Many large cities were built along its banks.

Starting Somewhere

Rivers have to get their start somewhere. For most, that's high up in the hills and mountains.

The start of a river is called its source. A river's source can be falling rain, melting snow, a lake from which a stream flows, or a spring that bubbles up from underground. As the trickle runs downhill, it combines with other trickles, collecting more water from rain and streams along the way.

Some rivers start very small from underground mountain springs.

Some rivers may look calm, but the power of the Colorado River carved the Grand Canyon!

From its source, a river flows downhill. It moves quickly, and the water can be heavy and powerful. As the water rushes along, it picks up stones and soil from the bottom of the river. This causes erosion, allowing the river to cut riverbeds and V-shaped **valleys** into the land where it travels.

WORD WISE

VALLEYS ARE DEEP GROOVES IN THE LAND CARVED BY RIVERS.

Parts of a River

After coming down mountains and hills and moving on to flat land, rivers slow down. Other streams and rivers flow into the main river, making it wider. As the water begins to flow more slowly, it loses power. The river cuts less deeply into the land, and it cannot rush over obstacles such as areas of very hard rock. Instead, the river has to flow around them.

Some rivers have bends and loops as they wind through land. These are called meanders.

Here you can see how the River Etive in Scotland has dumped a lot of sand, soil, and rocks called sediment where it meets Loch Etive. Loch is what they call a lake in Scotland.

The mouth of a river is where it empties into a bigger river, a lake, or an ocean. Many towns and cities are built at river mouths because these are good places from which to trade and travel. The mouth is where everything that comes down a river ends up—sediment, pollution, and fish.

Consider This

Why do you think river mouths are home to big fish and other animals that eat small fish?

How Lakes Form

A lake is a large body of water that is surrounded by land. Water in a lake is still and does not flow. Lakes form where there is a hole in the land and water to fill it. Some lakes form in the large pits called craters at the top of old volcanoes. Rain falls and fills the space, forming a lake.

Some lakes were formed by glaciers. Glaciers are slowly moving rivers of ice. As giant glaciers moved, they carved out basins and valleys in the land. Later, the ice melted. Some of this water collected in the holes, forming lakes.

People dig out land to make lakes, too. These are called reservoirs. Reservoirs catch and store water that can be sent to towns and cities for people to use.

Crater Lake in Oregon formed when the top of Mount Mazama fell during a volcanic eruption.

compare and contrast

How are lakes and rivers similar? How are they different?

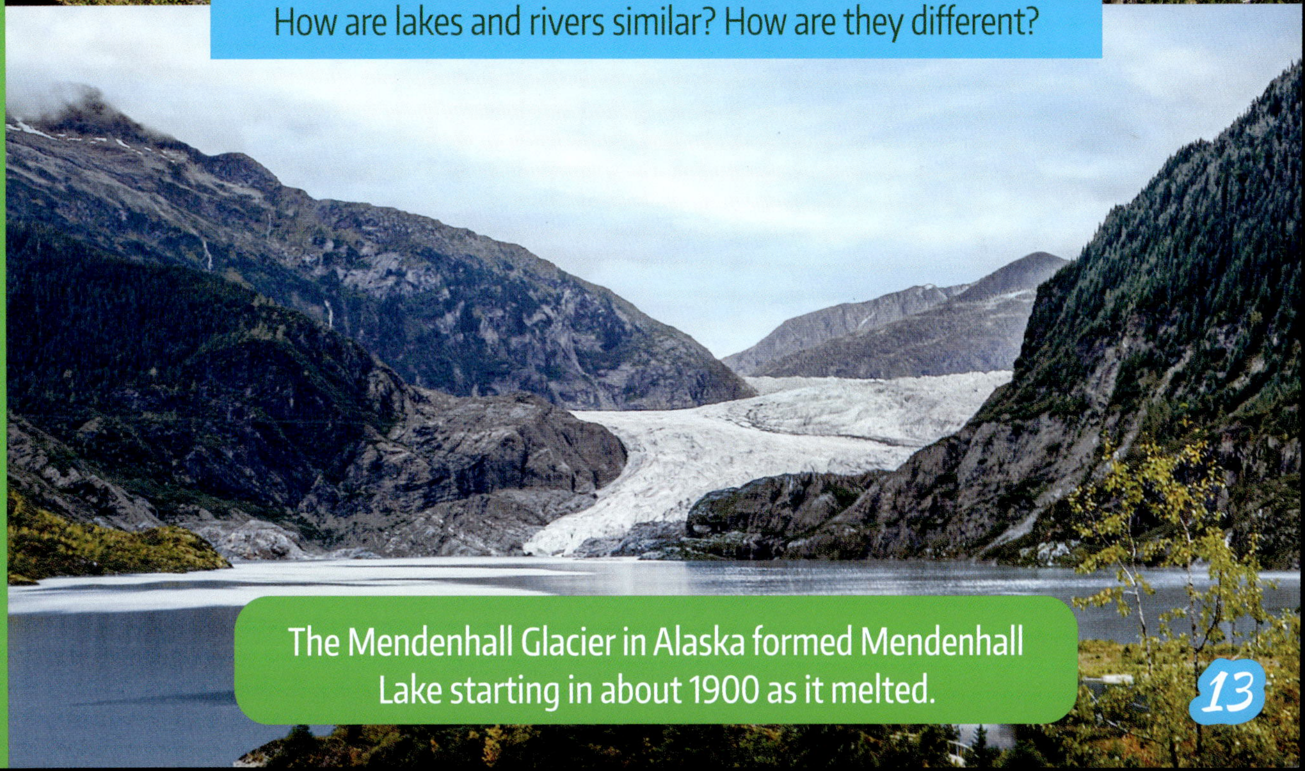

The Mendenhall Glacier in Alaska formed Mendenhall Lake starting in about 1900 as it melted.

Types of Lakes

Lakes can be different sizes; some seem as big as an ocean! The highest large lake in the world is Lake Titicaca in South America. It was formed by melted ice and snow filling a hollow in the Andes Mountains. Lake Baikal in Siberia is the deepest, oldest lake in the world.

Lake Baikal is in one of the coldest places on Earth, causing interesting ice shapes during the winter. Here, gas bubbles became trapped as the water froze solid.

Salt piles along the shore of the Great Salt Lake. The lake also has a pink color due to a special kind of algae that grows in the salt water.

Some lakes contain salty water. Lakes become salty if their water never drains away. River water that flows into these lakes is fresh but carries salts from the rocks it flows over. As water **evaporates** from the lake, the salts stay behind. Because of the high amount of salt in the Great Salt Lake in Utah, no fish live there.

WORD WISE

WHEN WATER EVAPORATES, IT BECOMES WARMER, DRIES UP, AND CHANGES INTO WATER VAPOR.

Disappearing Lakes

Lakes change all the time. In fact, most lakes last for only a few thousand years. They slowly disappear because when streams and rivers carry water into a lake, they bring sand, small rocks, and soil, too. This sediment settles on the bottom of the lake, and as it builds up, it gradually fills up the lake. This makes the lake shallower and shallower, until it disappears.

People change lakes, too. In some places, people take water from rivers to water plants on farms, to use in their homes, or to cool machines in big factories. This results in less water in the river to flow into lakes. This, too, can make a lake disappear.

Consider This

The world is getting warmer as a result of climate change. How might this affect lakes?

The amount of water in Lake Mead in the U.S. Southwest had dropped a great deal in recent years. This is because of drought and increased water use by people who live near it.

The World's Oceans

Oceans are much larger than lakes and all are filled with salt water. There are four main oceans in the world: the Pacific, the Indian, the Atlantic, and the Arctic. These oceans are joined, and water is always moving within and between them. One reason ocean water moves is that when strong winds blow across an ocean's surface, they create waves.

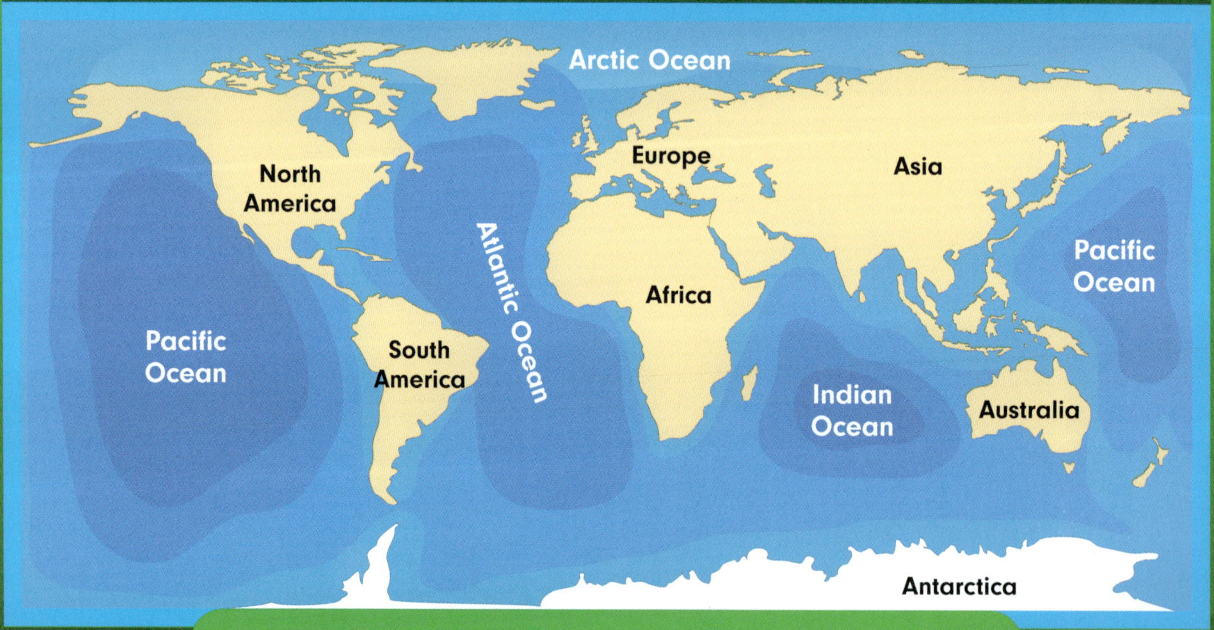

The Pacific Ocean, which lies between the Americas, and Asia and Australia, is the largest ocean in the world.

The ice in the Arctic Ocean is home to many animals such as polar bears, pictured, and seals.

Parts of the Arctic Ocean are always frozen into ice. The Arctic Ocean is the smallest, shallowest, and coldest ocean in the world. In the Arctic Ocean, there are huge blocks of floating ice up to 160 miles (257 km) wide!

Consider This

All the oceans are joined. Why do you think the world changed when people began to build oceangoing ships?

Where Rivers and Oceans Meet

Rivers supply the world's oceans with most of their water. Some water enters oceans by seeping up out of the ground or falling into oceans as rain. The place where fresh water from rivers meets and mixes with the salty water of the ocean is called an estuary.

The water in an estuary is a mixture of fresh water and salt water. This type of water is called brackish.

Deltas can change a lot over a short time. The Mississippi River Delta is often affected by hurricanes where it meets the Gulf of Mexico.

When rivers flow into the sea, they are moving slowly over flatter land. As they move more slowly, they drop sediment—loads of mud, rock, and sand. The sea may carry the sediment away, or it may settle and form a flat, muddy area called a delta. Sometimes, the river splits into small streams that flow around a buildup of sediment at a river's mouth. This is called a deltaic lobe.

Consider This

Rivers wash salt off rocks as they flow across the land. How does this explain why the sea is so salty?

Changing Tides

Just as rivers change all the time, so too does the level of water in the oceans. In most parts of the world, seawater gradually rises and then falls onto the shores where the oceans meet the land. These movements are called tides. Tides happen throughout the day. High tide is when the sea reaches its highest point on the coast. Low tide is when it has flowed all the way out again.

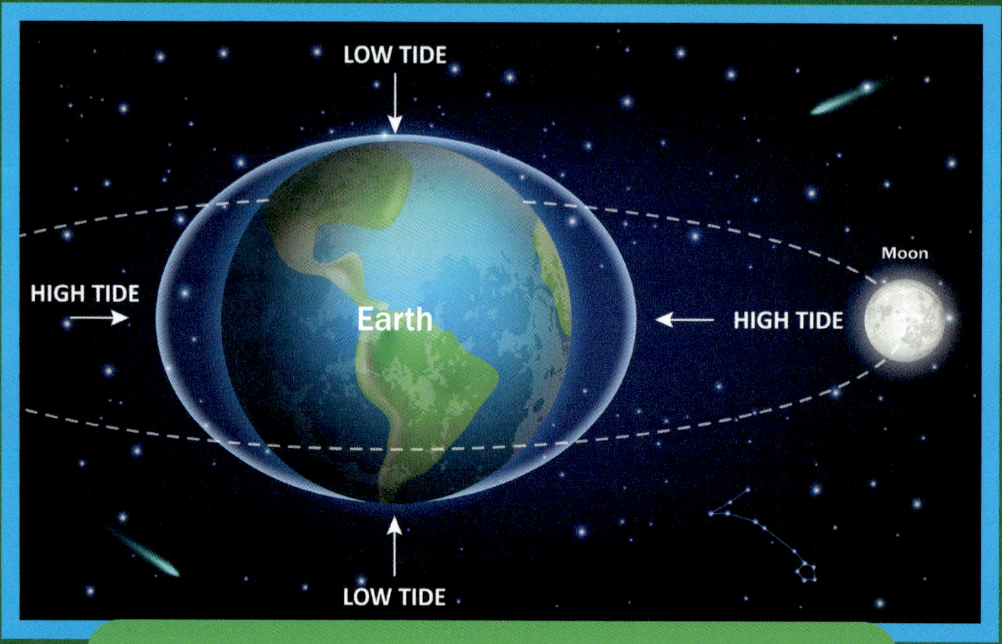

When the moon creates high tide on one side of the Earth, the opposite side also has high tide.

During low tide, some boats may end up sitting directly on the ocean floor. When the tide comes back in, they'll float again!

Tides happen because of the moon's **gravity**. Earth is spinning all the time. As it turns, the moon's gravity pulls at the oceans nearest to it. This makes the water rise and causes a high tide. Six hours later, Earth has turned 90 degrees. Now a different part of the Earth is facing the moon, and a different area of the world will have high tide.

WORD WISE

GRAVITY IS THE FORCE THAT PULLS ON THINGS.

Deep in the Ocean

Many animals and plants live in the ocean. Most of these living things are found near the water's surface. That is because the deeper the water, the darker and colder it is. At the surface, tiny plants use sunlight to make their own food. Tiny animals then eat these plants, and other sea animals eat them.

The closer to the surface of the water, the more sunlight can reach plants in the ocean.

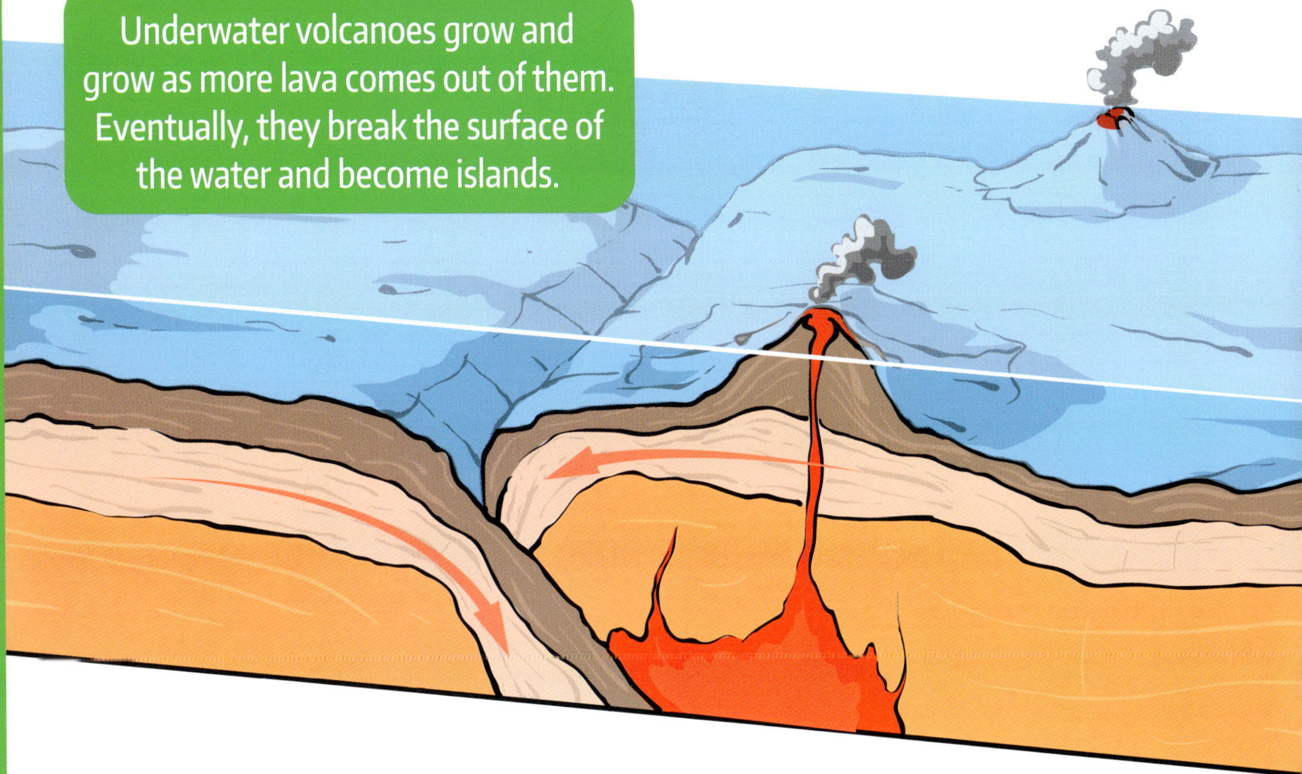

Underwater volcanoes grow and grow as more lava comes out of them. Eventually, they break the surface of the water and become islands.

The ground at the bottom of the ocean may be dark, but it is not completely flat. There are huge, flat areas of ground covered in sand, but there are also high mountains that jut up from the sea floor. There are deep valleys, or cracks, in the ground called trenches, too. In fact, the deepest valley in the world, the Mariana Trench, is found underwater!

Consider This

Why do you think it gets darker and colder the deeper you go in an ocean?

Water Pollution

Pollution is a major problem in rivers, lakes, and oceans all over the world. Farmers use sprays and powders on their fields to help crops grow and to kill bugs that harm the plants. Rain can wash these sprays into rivers and streams. In some places, dirty water from bathrooms and kitchens is flushed straight into rivers, lakes, and seas.

One reason it's so important to pick up your dog's waste during walks is because otherwise it'll end up in your city's water system.

Scientists have found plastic inside fish, turtles, and other sea animals.

Polluted water can be harmful to living things. Some animals die when they mistake litter for food. For example, turtles often mistake plastic bags in the sea for jellyfish and choke when they eat the bags. Polluted water is also unsafe for humans to drink or use for crops.

WORD WISE

POLLUTE MEANS TO ADD DIRTY OR HARMFUL SUBSTANCES TO AIR, SOIL, OR WATER.

Protecting Our Water

Water is one of the most important resources we have on Earth. That's why we need to protect rivers, lakes, and oceans. People lose at least 3 pints (1.4 l) of water every day through body processes, such as sweating. We must drink plenty of water to replace the water that we lose. We also use moving water in rivers to turn machines that make electricity.

The Hoover Dam protects the nearby areas from flooding, provides water for crops, and creates electricity.

We must protect the Earth's waterways so we can all have safe water to drink and play in.

We use water for washing, cooking, and cleaning. We catch fish and other foods from rivers, lakes, and oceans. We use water to grow plants. Farmers give water to animals to drink so they can provide us with milk, eggs, and meat. Humans also like to play in water, including swimming, boating, and diving.

Consider This

How important is water to life on Earth? What would happen without it?

Glossary

algae Simple plants that grow in or near water.
carve To cut away.
dehydrate To remove water from something. In humans, to become sick from lack of water.
drought A period of time during which there is very little or no rain.
erosion The wearing away of the Earth's surface by wind or water.
eruption In volcanoes, the act of exploding and sending lava and ash above the surface of the Earth.
fresh water Water that comes from rivers and lakes. Fresh water is not salty.
obstacles Things that are in the path of a river.
protect To keep safe.
resource Something that can be used.
source The start of a river.
transportation The act of moving items or people from one place to another.
urine Waste material that is secreted by the kidneys.
volcano A hole in Earth's surface from which hot melted rock called lava can spurt.
water vapor Water that is in the form of a gas in the air. Water comes in three states: water vapor (gas), water (liquid water), and ice (solid water).

For More Information

Books

Castaldo, Nancy F. *When the World Runs Dry: Earth's Water in Crisis*. Chapel Hill, NC: Algonquin Young Readers, 2022.

Kenney, Karen Latchana. *Oceans*. Minneapolis, MN: Bellwether Media, 2022.

Websites

Britannica Kids: Water Pollution
www.kids.britannica.com/students/article/water-pollution/598794
Read more about water pollution on the Britannica Kids website.

National Ocean Service for Kids
www.oceanservice.noaa.gov/kids
Learn all about the ocean in this National Ocean Service website with lots of photos and fun activities.

Publisher's note to educators and parents: Our editors have carefully reviewed these websites to ensure that they are suitable for students. Many websites change frequently, however, and we cannot guarantee that a site's future contents will continue to meet our high standards of quality and educational value. Be advised that students should be closely supervised whenever they access the internet.

Index